ROBOTICS

Tony Potter and Ivor Guild

Designed by **Roger Priddy**

Consultant Editor: **Nigel Ba** E.L.R.S.
Editor: **Lynn Myring**
Robot program: **Chris Oxlade**

Illustrated by: **Jeremy Gower**
**Chris Lyon, Simon Roulstone, Martin Newton,
Geoff Dicks, Mick Gillah, Rob McCaig,
Tim Cowdell, Janos Marffy, Mike Saunders,
Kuo Kang Chen, Stan North.**

CONTENTS

3 **About robotics**

4 **What robots can and cannot do**

6 **Mobile robots**

8 **How arm robots work**

10 **Designing robots**

12 **Special purpose robots**

14 **Robots in space**

16 **Micro-robots**

18 **Robot factory workers**

20 **How to teach a robot**

22 **Types of arm robot**

24 **How robots are driven**

26 **How robots hold things**

28 **Computer control**

30 **Sensors**

32 **How a robot knows where it is**

34 **Cybernetics**

36 **Latest developments**

38 **Build your own micro-robot**

47 **Robot words**

48 **Index**

First published in 1983
by Usborne Publishing Ltd, 20 Garrick Street, London WC2E 9BJ

© 1983 by Usborne Publishing Ltd

About robotics

Seventy years ago no one had ever heard the word "robot". It was first used by a Czechoslovakian writer, Karel Capek (pronounced Chapek) in the 1920s. He wrote a play about a scientist who invents machines which he calls robots, from the Czech word *robota*, meaning "slave-like work". He gave them this name because they were used to do very boring work. At the end of the play, the robots kill their human owners and take over the world.

There are many robots in existence now, but they are quite different from the robots of science fiction films and books. Instead of being frightening, super-intelligent metal

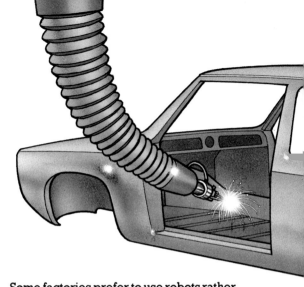

Some factories prefer to use robots rather than other automatic machines because they can be re-programmed to do different jobs.

people, real robots are just machines controlled by a computer to work in a set way. They are generally deaf, dumb, blind, have no sense of taste, smell or touch, have difficulty getting around, and have no intelligence of their own. However, advances in microchip technology mean that robots are beginning to be made with sensors – a TV camera "eye" or a microphone "ear" for instance – which give them very limited senses like electronic sight and hearing.

Robots are used to do many things, often jobs which are very dangerous or tiring for people to do, like welding car bodies. In factories, robots are useful because they are often able to work more efficiently than people. Although robots break down, they never need holidays, sleep or meal breaks.

Some robots are used to do jobs that would be impossible for people to do, such as working inside the radioactive part of a nuclear power station, or visiting distant planets. Others, like the small micro-robots used with a home computer, are just for fun or for learning about robotics. You can find out how to make your own micro-robot on page 38.

What robots can and cannot do

Robots are able to do many different things, especially in factories. Here robots are carefully maintained and organized to work alongside other automatic machines. Robots are rarely used to work outside because it is much more difficult to get them to work away from the ordered environment of a factory.

Science fiction robots are often made to look human, but the appearance and ability of an industrial robot depend on the kind of work it has to do. The majority of them are like "arms" bolted to the floor because the work they do can be done standing in one place. These arm robots are often called manipulative robots because they hold things – perhaps a tool like the welding torch in the picture below.

Arm robots are most familiar in car factories, but they can be found in many other industries – electronics, engineering, clothing and confectionery, for example. The jobs they are best at are those that involve doing the same thing over and over again.

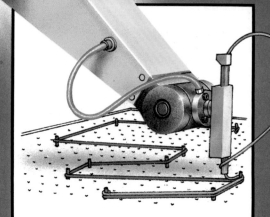

Accurate robots

Some robots are able to do very accurate and intricate work, but this depends on the design of the robot and the computer program that controls it. This picture shows a robot laying out lengths of wire in a complex pattern for wiring-up electric vehicles, like fork-lift trucks. To do this, it first has to push nails into holes in a pegboard according to a pattern stored in its computer memory. The robot has to line up the nails with the holes very precisely to get them to fit.

Robots today

Most of today's robots are only able to work in a factory where everything is carefully organized around them. Robots are usually next to a conveyor belt which "feeds" them with work, for example. They also have to be kept in a wire "cage" to prevent the robot from injuring any passing people. Some scientists believe that in fifty years or so it will be possible to build a robot capable of working anywhere. This kind of robot would have to be much more "intelligent" than existing robots and have lots of sensors to be able to react to a vast amount of information, or feedback, about the world around it. Even the best of today's robots could not react fast enough to catch a ball, for instance. Imagine trying to do this with thick mittens on, one hand tied behind your back, your eyes blindfolded, feet cemented to the ground and your ears and nose blocked up. Most robots have to rely on even less feedback than this.

Tough robots

Many robots can do work which would be dangerous or unpleasant for people. Robots are very tough because they are made of metal, and can withstand very extreme conditions, such as a hot poisonous atmosphere. This robot is putting its hand right into a red-hot oven to take a metal casting out. The heat does not affect its performance, so it is able to produce high quality castings by always taking them out at the right temperature.

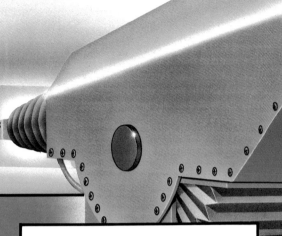

What happens when things go wrong?

Many robots are unable to react to anything unexpected happening to them because they have no sensors. This robot, pictured below, is controlled by a computer to spray bicycle frames as they pass by on a conveyor. If a frame falls off, the robot carries on spraying. One way of preventing this is to put a switch on the conveyor which turns off the robot. Another way to stop it is to give the robot electronic senses to detect what is going on.

How strong is a robot?

The strength of a robot depends on the power of its motors and the materials it is made from. A home micro-robot made from thin sheet metal can only lift the weight of an apple, for example. But a large industrial robot like the one above could pick up something as heavy as an elephant. A robot like this can easily lift sacks all day, whereas even a strong person would get tired eventually.

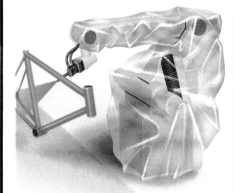

This robot wears a plastic cover as an "overall" to stop paint clogging it up. Other robots also need to wear special covers.

Mobile robots

A mobile robot is a computer-controlled vehicle of some kind, and the most common have wheels or tracks. Some carry a computer around with them, but others are connected to a computer by a long cable or by radio link. Mobile robots are beginning to be used in factories to move goods and materials around, sometimes from one arm robot to another. Robot trucks like the one in the picture below can be guided round the factory by following white lines or magnetic signals from cables buried underground. The computer controlling the truck is programmed to tell it which route to take round the network of lines or cables.

Steering a robot

This is a micro-robot called Bigtrak. It is steered by two motors driving the wheels in the centre. Its computer controls the steering by changing the speed and direction of the motors. Bigtrak's other wheels just prevent it tipping up.

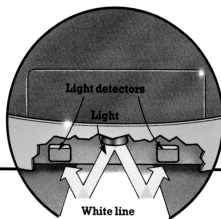

Light detectors

Light

White line

How robots follow buried cables

How robots follow lines

The circle above shows how a robot can follow lines by using a sensor to send feedback to the computer about its position over the line. The sensor usually consists of a light pointing to the floor with a light detector either side of it. These detect light reflected from the line. If the robot steers off the line, they send a message to the computer to correct the steering.

Bumper   Sensor

Cable

Magnetic field

A cable-following robot like the one above usually has two coils of wire fixed to the front which detect a magnetic field surrounding the buried cable. The magnetic field is made by sending electricity through the cable and this field, in turn,

creates a small electric current in the robot's coils. The strength of the current in the coils alters according to how far the robot is from the cable. The computer steers the robot over the cable by balancing the strength of current in the two coils.

Forwards or backwards: Both wheels are driven at the same speed in the same direction.

Right turn: The left-hand wheel is driven forwards, and the right-hand wheel backwards.

Left turn: The right-hand wheel is driven forwards, and the left-hand wheel backwards.

Robots with tracks can be steered like this too because each track is like a driven wheel.

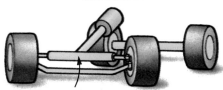

Steering mechanism

This kind of robot steers by turning on the spot. Robots can also be steered gradually as they move forward by making both motors go forward, with one going faster.

Some robots use a steering mechanism like that in a car. Instead of a steering wheel they have a motor connected to a computer. Robots like this are less manoeuvreable than the Bigtrak type because they cannot turn on the spot.

Roving robot

New mobile robots are being developed which find their way around by navigation. This means that the robot's computer has to decide how to get the robot from where it is to where it wants to go without following any guides, like white lines, and without bumping into anything.

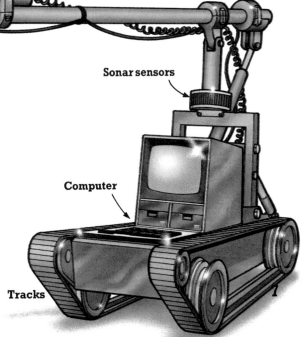

Grab arm

Sonar sensors

Computer

Tracks

This picture shows an experimental robot called Mr Bill (Mr stands for Mobile Robot) which navigates by using sensors to find out about its environment. Most of the information comes from sonar sensors mounted on the robot. These work by emitting a sound, and then "listening" for an echo to bounce back from obstacles. This information is compared with a "map" of the positions of fixed obstacles, like walls, which is stored in the memory of the on-board computer. There are also other sensors on the robot's wheels, called odometers, which tell the computer how far the robot has travelled. The computer works out where the robot is by making calculations using all this information.

How arm robots work

These pages show all the things that are needed to make an arm robot work and a view inside the robot to show what makes it move. There are many different ways of constructing and powering a robot. This one is jointed like a human arm and is driven by electric motors. Some industrial robots work in the same way, but are much more complex inside.

Each moving part of a robot is usually driven by a separate source of power. This one has six motors to make its arm and wrist parts move. Each motor is switched on and off by the computer, which also controls the speed of the motors. You can find out about other ways of driving robots on page 24.

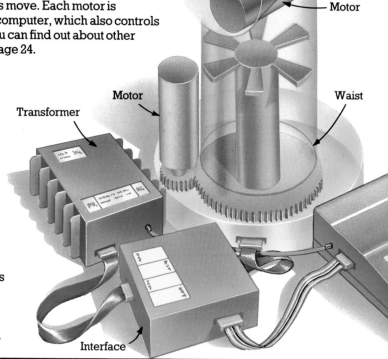

Shoulder

Gear

Motor

Transformer

Electricity for the robot's motors and computer is provided by a transformer. This converts the strong electric voltage from the mains into a low voltage

Transformer

Motor

Waist

Interface

A device called the interface links the transformer, motors and computer together. Inside is an electronic circuit which switches the power to the motors on and off when instructed by the computer.

Interface

Arm movements

Waist

Shoulder

Elbow

Most arm robots have three main parts which are joined together. The point where one part is fixed to another is usually called a joint or axis. The joints on a robot like the one above are given the names elbow, shoulder and waist. Each axis is said to give the robot one degree of freedom because it allows the parts fixed at the joint to move in a certain way. In these pictures you can see the direction in which each joint allows the robot to move.

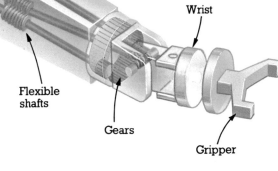

There are three tiny motors inside the robot's "forearm" which drive three moving parts in the wrist. These motors are connected to gears in the wrist by very long shafts. Each shaft has a flexible joint in the middle which allows it to bend as the wrist moves from side to side and up and down.

Elbow

Motors

Wrist

Each motor is connected by gears to a shaft which moves part of the robot. In this picture the shafts are the parts painted orange, and the gears are painted green. The gears help to reduce the speed of the motors.

Flexible shafts

Gears

Gripper

The computer is programmed using the keyboard. It controls everything the robot does by sending a sequence of instructions to the interface.

The robot's hand, called a gripper, is shown separated from its wrist in this picture. You can find out how these work on page 26. The wrist is a complicated mechanism which can bend in three ways, shown in the picture below. Some robots have wrists which bend in only two directions, but this depends on the kind of work they have to do. The more joints in the wrist, the more able the robot is to make fine movements to do a job.

Wrist movements

Roll

Pitch

Yaw

Between the gripper and the end of the robot arm is a kind of wrist. Like the arm, the wrist usually has three joints, or axes of rotation. These allow the gripper to move in three ways, shown in the pictures above. These movements have special names: yaw, pitch and roll. A robot like this which can make six kinds of movements has six degrees of freedom. Some robots have more than this, some less, depending on the kind of work they do.

Designing robots

It is very difficult to design and build a robot, even to do a simple job. A robot designer has to begin by breaking down the job into as many steps as possible to see what sort of robot is needed. For example, the robot arm below would need to be able to bend its wrist if it had to lift a glass of water. These pages shown an imaginary robot servant designed to do all the dusting in a two-storey house. An extremely complex robot is needed to do this apparently simple task. Experts think it may be possible to build a robot like this in a few years time.

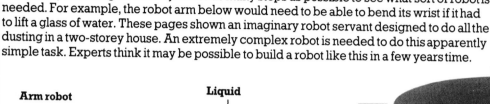

Arm robot

Liquid

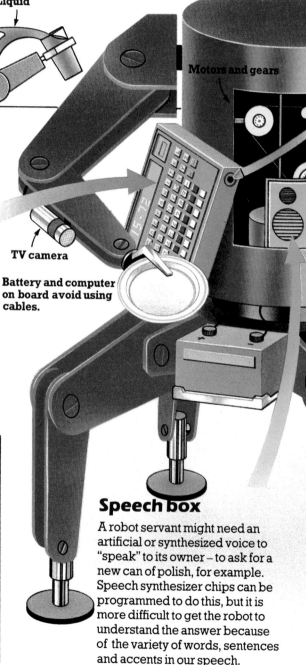

Motors and gears

TV camera

Battery and computer on board avoid using cables.

Computer control

The computer has to be programmed to control everything the robot does: how the motors drive its legs and arms, how it navigates around the house without causing damage, the way it does its dusting, and so on. The computer must make the robot do everything in the right order, such as opening doors before going through them. It also has to work things out without delay so that the robot can respond instantly to something unexpected, like a baby crawling under its feet.

The program

An extremely complex program would be needed for the computer because the robot's job involves hundreds of choices based on information, or data, about the world around it. This part of the design, called the software, gives the robot "intelligence" so that it can "decide" what to do.

Speech box

A robot servant might need an artificial or synthesized voice to "speak" to its owner – to ask for a new can of polish, for example. Speech synthesizer chips can be programmed to do this, but it is more difficult to get the robot to understand the answer because of the variety of words, sentences and accents in our speech.

Arms

The robot has two arms because it needs to be able to hold things while dusting underneath. It could also have a spray-can of polish, with a computer-controlled plunger to press the button, fixed to one arm. This would avoid the need for a third arm to hold the can.

Can of polish

<table>
<tr><td>

Design your own robot

Try drawing a robot to do one of these jobs: **1** Take a dog for a walk. **2** Wash the dishes. **3** Mend a puncture in a bicycle tyre.

</td></tr>
</table>

Sensors

Different kinds of sensors are required for the robot to do its job: navigation sensors to find its way around, TV camera "eyes" to so that it can "see" what is doing, and safety touch sensors which stop it if it accidently bumps into anything. All the data from the sensors is sent to the computer via an interface so that it can control the robot's actions.

Legs

This robot needs at least four legs to climb the steep stairs of a house – something it could not do with wheels or tracks. A Japanese designer has actually built a four-legged, stair climbing robot.

Walking robots

This shows what happens when a robot is built with different numbers of legs.

One leg

A one-legged robot like this has to keep hopping to balance, so it would not be much good for a job like dusting.

Two legs

When a two-legged robot takes one foot off the ground to walk, it has to balance on one foot. This is very difficult for the computer to control.

Three legs

A three-legged robot is very stable standing still, but as soon as it takes one foot off the ground to walk it falls over.

Four legs

A four-legged robot walks by moving one leg at a time. This means it always has three legs on the ground to balance with.

11

Special purpose robots

These pages show robots which have been specially designed by carefully working out all the things they need to be able to do for particular jobs. Sometimes factory arm robots can be adapted, but other jobs may need a completely new kind of design.

Sheep shearer

This is an experimental robot specially designed to shear sheep. The sheep is held down with straps on a cradle and shorn with electric clippers. The robot's computer gets feedback from sensors on the clippers so that it is able to position them just above the sheep's skin. If the sheep wriggles it can react in less than one ten-thousandth of a second to move the clippers away. An electronic "map" of the sheep's shape is stored in the computer's memory so that it can tell the robot where to cut.

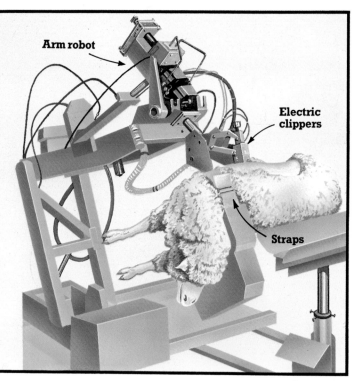

Arm robot

Electric clippers

Straps

Robot patient

This robot patient is designed to respond to treatment by students and can even "die" if someone makes a mistake. Computer-controlled electronic components inside the robot can be programmed to mimic breathing, heart-beat and blood pressure. Sensors inside the body measure the efficiency of a student's treatment.

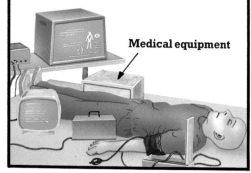

Medical equipment

Robot hand

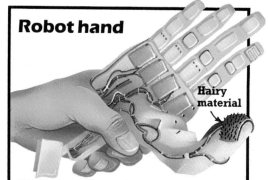

Hairy material

The picture above shows an experimental microchip-controlled false hand, which is activated by muscles in the wearer's arm. There is a microphone in the thumb covered by a strip of hairy material. The microphone "listens" for the rustling sound the material makes when an object is held. As the hairs are crushed, the sound stops, which tells the computer the grip is tight enough.

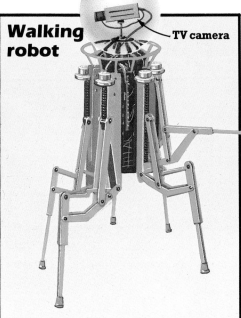

Walking robot

TV camera

This robot is able to walk over rough ground and go up stairs by adjusting the length of its legs. Inside the plastic dome at the top is a TV camera which sends pictures to a computer in the middle of the robot. The robot can walk around for about an hour before its batteries go flat.

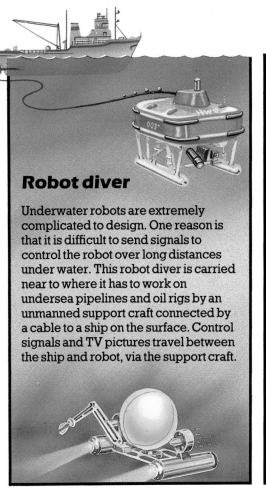

Robot diver

Underwater robots are extremely complicated to design. One reason is that it is difficult to send signals to control the robot over long distances under water. This robot diver is carried near to where it has to work on undersea pipelines and oil rigs by an unmanned support craft connected by a cable to a ship on the surface. Control signals and TV pictures travel between the ship and robot, via the support craft.

Robotics teacher

Hero 1 is a robot designed to teach people at school or in industry about robotics. It is a mobile and arm robot combined and has lots of different sensors so that students can discover what they are and how they work. It also has a voice synthesizer which can be programmed with its built-in computer.

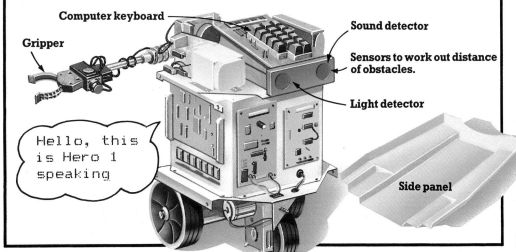

Computer keyboard

Gripper

Sound detector

Sensors to work out distance of obstacles.

Light detector

Hello, this is Hero 1 speaking

Side panel

Robots are particularly useful for doing jobs in space because it is such a hostile environment for humans to work in. In the future, robots and other automatic machines may make up most of the space workforce.

Space arm robot

The Space Shuttle can be fitted with a long, folding arm robot as part of its equipment. This is used for launching satellites and other machines from its cargo hold, or retrieving them from space for repair. The arm folds neatly out of the way in the cargo hold after use.

Cargo

TV camera

Elbow joint

Shoulder joint

Cargo hold

Shiny blanket

The arm, called RMS (Remote Manipulator System), has its own computer which is programmed to make 20 different sets of movements. It can also be controlled from the flight deck with joysticks similar to those used with computer games. Up to eight cameras can be positioned on the arm so that the operator can see what to do.

The RMS is capable of lifting an object which would weigh about the same as fifteen cars on Earth. It is designed to cope with twice as much in an emergency. If the arm gets stuck and prevents the cargo doors from closing it can be jettisoned into space.

A shiny blanket covers the whole arm so that it reflects the Sun's heat and does not get too hot. There are also heating elements inside the blanket to keep the arm warm when the Shuttle is on the night side of Earth.

Each joint is driven by a tiny electric motor. Sensors on the joints tell the computer the position of the arm.

TV camera

Wrist joint

Satellites

Satellites often include components like sensors and computers, but are really automatic machines rather than robots. The sensors on satellites are often used to collect data rather than to provide feedback for its computer.

Stretched out, the arm can reach nearly as far as the length of two buses put together.

Robot missiles

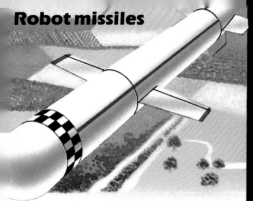

Some kinds of missiles are described by experts as robots. They are programmed to reach a target automatically using sensors and an on-board computer. Cruise missiles, for example, use sensors to "see" the ground below, and then compare this data with a computerized route map. This enables them to fly very low to avoid radar detection.

Space probes

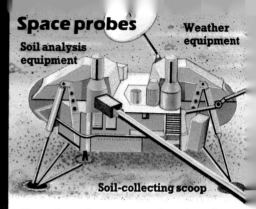

Soil analysis equipment

Weather equipment

Soil-collecting scoop

Apart from the moon, exploration of other planets in the solar system and beyond has only been made by robot spacecraft. This is mostly because of the time it takes to reach them – Voyager 1 took 18 months to reach Jupiter, for example. This picture shows a computer-controlled robot landing craft sent to the surface of Mars by Viking 1.

How the arm picks things up

The end of the arm has a special gripping mechanism inside, made of diagonally crossed wires, to hold satellites and other cargo. Each piece of cargo has a shaft sticking out of one end. The end of the arm is manoeuvred over the shaft, and then rotated. This twists the wires around the shaft so that the cargo is pulled tightly against the end of the arm. The end of the arm is simply rotated in the opposite direction to release the shaft.

The picture below shows the arm releasing a telecommunications satellite into orbit above the Earth.

Solar panels

Shaft

End gripper

Micro-robots

A micro-robot is a small robot controlled by a home computer. You can find out how to make your own micro-robot on page 38.

Drawing robot

The Turtle is a mobile robot which can be programmed to draw with a pen as it moves around. A computer language called LOGO makes the robot move in units of about 1.5mm at a time. LOGO uses commands like "F 10" for forward 10 units, or "R 90" for right 90 degrees, to draw simple shapes like squares or triangles. The commands are used to combine a series of simple shapes to make pictures.

Each unit the Turtle moves is measured by a sensor mounted over a cog on each wheel. A tiny lamp on one side shines a beam of light between the teeth of the cog into a photoelectric cell on the other side. The teeth break the beam of light as the wheel rotates. Each break in the light path is detected by the photoelectric cell, which sends a message to the computer to count one unit.

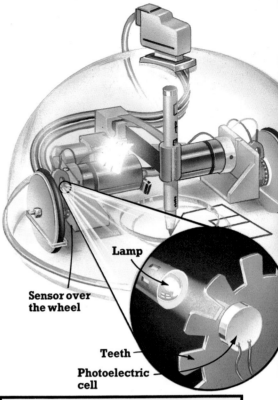

Lamp

Sensor over the wheel

Teeth

Photoelectric cell

How to connect up a micro-robot

These pictures show how a micro-robot arm is connected to a home computer and to a power supply.

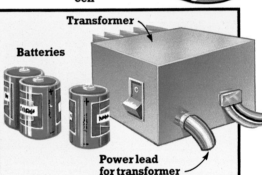

Transformer

Batteries

Power lead for transformer

Most micro-robots are driven by low voltage electric motors, so a transformer or batteries are used to power them. This power supply is usually connected to an electronic interface.

Port

Wires, known as control lines, which control the robot's motors, are plugged into a socket in the computer, called a port. (Not all home computers have the right ports to be able to do this.) One wire is often used for each motor.

It is very dangerous to plug a micro-robot into mains electricity.

Bar codes laid out in a long line.

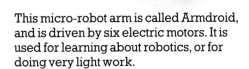

Draw like the Turtle

Try drawing the shape these LOGO commands make. Use 1mm units for the forward movements. F100, R90, F100, R90, F100, R90, F100.

Gears to move arm

Motors

The interface is connected to the robot's motors. It is made of electronic components which switch the power to the motors on and off when the computer sends signals to them. Sometimes this circuit is in a separate box, but it can be inside the robot or the computer.

This micro-robot arm is called Armdroid, and is driven by six electric motors. It is used for learning about robotics, or for doing very light work.

Buggy

The robot on the left is called the BBC Buggy. It is made from Fischertechnic construction kit parts, so it can be added to by building extra bits, like an arm, on top.

A sensor, called an infra-red transceiver, is fixed to the front. This works by transmitting invisible infra-red light down to the ground, and then receiving it back when it is reflected by the surface the robot travels over.

The computer can be programmed to use the data from the infra-red transceiver to "see" a line, and tell the robot to follow it, or "read" a bar code like the one in the picture. The computer translates the bar code into musical notes and can play a tune by going over a series of codes laid on the floor. Different kinds of sensors can be plugged into the front of the circuit board on top of the Buggy. Bumpers make the robot reverse automatically when it bumps into something.

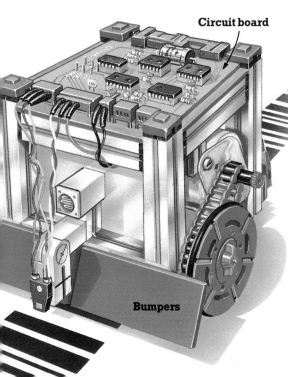

Circuit board

Bumpers

17

Robot factory workers

It is likely that very soon there will be almost totally unmanned factories. Perhaps just one or two people will program or monitor the computers and carry out routine maintenance to robots and other machines.

Car plants are currently among the most highly automated factories in the world. This picture shows how robots and other automatic machines, like conveyor belts and stackers, are used alongside each other to assemble and manufacture parts for cars.

Welding station

The framework built over the conveyor, shown on the right is called a welding station. It has six robots fixed to it holding welding guns. As car body panels, which have been lightly tacked together elsewhere in the factory, pass below, the robots weld them together to make a tough, rigid car bodies. As there are six robots working together, they can assemble cars very quickly.

Machining centre

The robots below are part of a system called a machining centre, or cell. One robot unloads heavy lumps of steel ready for the other robot which "serves" the two automatic lathes. A computer is in charge of the computers controlling the robots, lathes and conveyors to make sure each machine does the right thing at the right time. This is very important because otherwise the robots could collide, or damage the lathes.

Computers in metal cabinets

Welding station

Lathe

Finished parts

Conveyor belts

Lathe

Pallet

Steel

The computer-controlled lathes could be programmed to make many different parts – for gearboxes, axles, engines and so on. The "serving" robot loads the raw steel into the lathes and then unloads the finished part onto the conveyor for assembly or finishing in another part of the factory.

The robot on the left unloads the steel from an automatic shuttle, which is like a tiny flat truck on rails. The truck carries the steel on a pallet – a wooden or metal platform used to stack materials for transport.

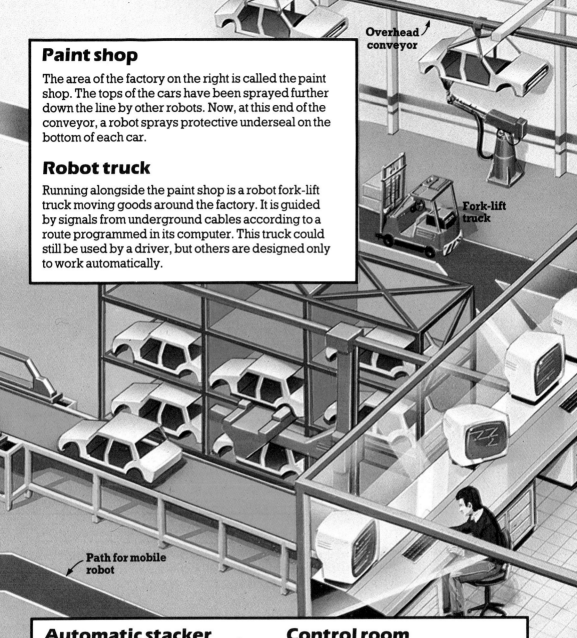

Paint shop

The area of the factory on the right is called the paint shop. The tops of the cars have been sprayed further down the line by other robots. Now, at this end of the conveyor, a robot sprays protective underseal on the bottom of each car.

Robot truck

Running alongside the paint shop is a robot fork-lift truck moving goods around the factory. It is guided by signals from underground cables according to a route programmed in its computer. This truck could still be used by a driver, but others are designed only to work automatically.

Overhead conveyor

Fork-lift truck

Path for mobile robot

Automatic stacker

The orange machine above is called an automatic stacker. It is programmed to place partly completed cars in a rack until they are needed. It saves floor space because it stacks things vertically. The same kind of computer-controlled stacker is used in Japanese cities to park bicycles. Some experts say these are robots because they can be programmed to stack different things, but others disagree.

Control room

Above is the control room where all the automatic operations being carried out in the factory are controlled and monitored. The computers here organize all the separate computers controlling the robots and other machinery. Someone watches display screens to check that all the machines are working properly and that production targets are met. Systems like this are already in use.

How to teach a robot

A robot's computer has to be given a set of instructions called a program to get it to work. This is done either by guiding the robot through a sequence of movements, and programming the computer to remember them, or by instructing the computer directly with the keyboard. In this way, the robot can be made to "learn" a set of movements, and repeat them over and over again.

Sensors

Showing an arm robot what to do

One way of teaching a robot is by guiding its arm through the movements needed to do a job. This is called lead-through programming. The robot in this picture is being taught to spray paint by a person skilled at the job.

First the computer is programmed to remember the movements shown to the robot, and the order they were made in. Then the computer is programmed to make the robot automatically follow the path it was taught. It is very important that the computer repeats this exactly, otherwise the robot would spray paint in the wrong pattern and make a mess. Sensors on joints send data to computer about robot's position.

Remote teaching

Robots can be taught remotely with a computer keyboard, or a simplified keyboard called a teach pendant. This is connected to the computer and has commands like UP, DOWN, LEFT and RIGHT, which can be used to manouevre the robot. It also has a TEACH button which is pressed to make the computer to remember positions the operator wants it to know.

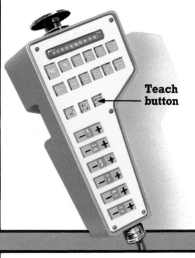

Teach button

Direct control

Forward 10
Left 90
Forward 24
Left 45
Forward 35

This micro-robot is called the Zeaker. It can be programmed to move around, using a computer language with similar commands to LOGO. It can also be used for drawing as there is a pen fixed under its body.

Talking to robots

Micro-arm robot

← **Interface**

Speech control can be used by disabled people unable to use a keypad.

Microphone

People are experimenting with controlling robots by giving them spoken instructions. This is done by connecting a microphone to the computer via a special interface. The interface converts commands such as "up" or "down" into a sequence of electrical signals which the computer is programmed to recognize as instructions for the robot.

The program, or software, which comes with the Zeaker lets the user direct its movements by pressing keys on the computer. A sequence of movements can be built up in the computer's memory and repeated over and over again to get the robot to draw complex patterns.

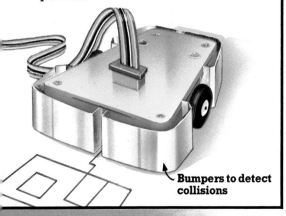

Bumpers to detect collisions

Chocolate-box packer

For some jobs, like picking things up and putting them down, the robot only needs to know precisely the points to start and finish at. The robot can be shown what to do by being guided to these points, by hand or with a teach pendant, and then getting its computer to remember them. The computer is programmed to work out the route the robot takes between the points. Industrial robots are taught to do loading and simple assembly jobs in this way.

In a factory the box would be on a conveyor.

Conveyor

Step 1: The robot is guided to the chocolate and its computer told to remember the point where it has to open and close its gripper.

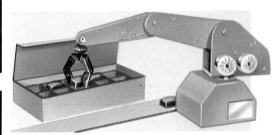

Step 2: The robot is guided to the point where it has to drop the chocolate into the box, and the computer instructed to remember it.

Step 3: The stopping and starting points are now in the computer's memory and it can tell the robot to repeat the movement over and over again.

Types of arm robot

There are five main types of arm robot, each designed to be able to move in different ways according to how its moving parts are put together. The design of a robot is called its architecture, and the space it can move around in as a result of its design is called its working envelope – shaded blue on these pages.

Jointed-arm robots

The design of a jointed-arm robot is based on the human arm. The one on the right has a rotating base part which is not able to go all the way round. The arm is jointed at the shoulder and elbow and can bend like a door hinge at both joints. The working envelope of a jointed-arm robot is shaped like part of a ball.

Spherical robot

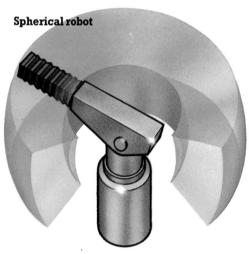

Spherical or Polar robots

This type of robot gets its name from the spherical working envelope it is able to move in. The main arm part of the robot on the left moves in and out like a telescope, and also has a hinge-like joint at the shoulder. The robot's waist rotates, but it cannot go round 360°. The design of spherical robots makes them very strong, so they are often used to pick up heavy weights – sometimes as much as the weight of a car.

XYZ robot

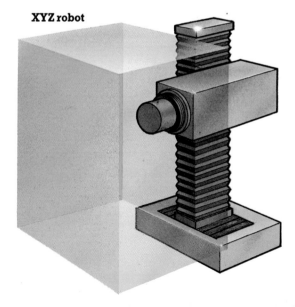

XYZ robots

Robots like the one on the right get their name because they are able to move in three different directions called X, Y and Z and have a cube-shaped working envelope. The robot's side to side movement on its base is called direction X. The main arm part goes in and out telescopically, and this is direction Y. This part of the arm also moves up and down in direction Z. The design of XYZ robots makes them very accurate, so they are often used to do precise jobs like assembling things.

22

Cylindrical robots

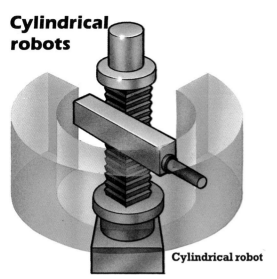

Cylindrical robot

The main arm part of a cylindrical robot moves in and out telescopically, and is also fixed to a "pole" at the shoulder so that it can slide up and down. The "pole" rotates, although not all the way round, and this gives the robot a working envelope shaped like a cylinder.

Spine robot

This is a new type of robot designed on the same principle as the human spine. The Spine robot can reach almost anywhere within its working envelope, even back into the centre, so it can work in inaccessible spaces like the inside of a car. The arm can also swing right round in a circle over and over again.

Inside its concertina cover are lots of discs piled on top of each other. The robot can be made longer or shorter by adding or removing discs. The discs are held together by two pairs of cables which are fixed to pistons in the base. The robot's computer controls the spine by moving the pistons to pull on the cables.

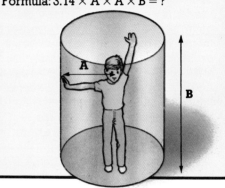

Your working envelope

Try working out the volume of your own working envelope by imagining that you are standing inside a cylinder with one arm stretched out to the side and the other straight above you. Get a friend to make the measurements shown in the picture and put them in place of the letters in the formula. Your answer should be in cubic centimetres or cubic inches, depending on which kind of measurements you use.

Formula: $3.14 \times A \times A \times B = ?$

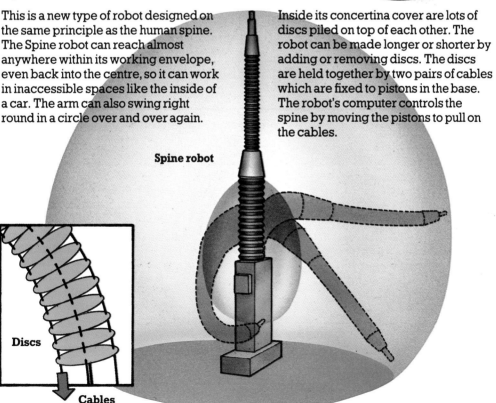

Spine robot

Discs

Cables

How robots are driven

Each moving part of a robot is driven separately, either by an electric motor or by a hydraulic or pneumatic system. The speed of the drive must be able to be varied so that the robot can be controlled to move quickly or slowly. Mobile robots are usually electrically driven, but the kind of drive used on an arm robot depends on the work it has to do.

Electric motors

Many different types of electric motor are used to drive robots. One type, which is often used, is called a direct current or d.c. motor. The picture below shows how a simplified d.c. motor looks inside. The gear at the bottom of the shaft that goes through the centre of the motor would be connected to part of a robot to drive it.

On either side of the motor are permanent magnets, one with a north and the other with a south pole facing the centre. Electricity is passed through the brown contact on the right, round the wire coil, out through the contact on the left and back to the battery. This makes the coil an electromagnet, with a north pole on one side (shown in green) and a south pole on the other (shown in yellow).

Hydraulic systems

A hydraulic system works rather like a syringe used for injections. It can be used to make either circular or straight movements according to the type.

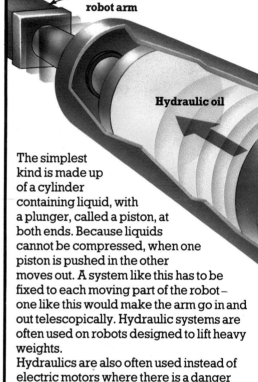

End of robot arm

Hydraulic oil

The simplest kind is made up of a cylinder containing liquid, with a plunger, called a piston, at both ends. Because liquids cannot be compressed, when one piston is pushed in the other moves out. A system like this has to be fixed to each moving part of the robot – one like this would make the arm go in and out telescopically. Hydraulic systems are often used on robots designed to lift heavy weights.

Hydraulics are also often used instead of electric motors where there is a danger that sparks from a motor might ignite fumes in the atmosphere of a factory.

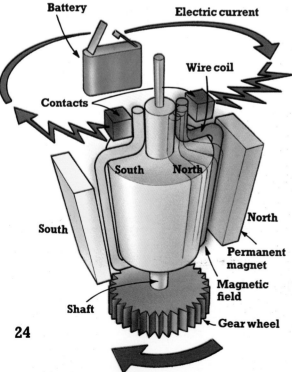

Battery

Electric current

Wire coil

Contacts

South **North**

North

South

Permanent magnet

Magnetic field

Shaft

Gear wheel

Because these poles are the same as the permanent magnets opposite them, the wire coil is repelled. When the coil is forced round half a turn, the parts touching the contacts will have changed positions, which again gives the coil a north pole on the right and a south pole on the left. The permanent magnets repel the coil again, and the whole process is repeated, over and over again.

The speed of a motor like this can be reduced or increased using gears. The speed can also be varied electrically with a device similar to the foot control on an electric sewing machine. The strength of a robot partly depends on the speed of its motors – usually the slower the motor the more power it has.

There are different ways of pushing the pistons in and out on a hydraulic system – the type used on car brakes, for example, is operated by a person pressing the brake pedal. On a robot, however, the piston has to be pushed by some electrically operated device for the computer to be able to control it. This picture shows a device called a solenoid fixed to a shaft on the end of the piston. This too has a piston inside which is pushed in and out by an electromagnet. You can see how these work on page 27

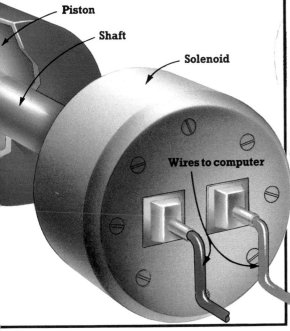

Piston

Shaft

Solenoid

Wires to computer

Pneumatic systems

In a pneumatic system, air, or some other gas, is used to move a mechanical part of the robot, often the gripper. A simple system consists of a cylinder with a piston inside which is connected to a shaft on the robot. Compressed air is let into one end of the cylinder by a computer-controlled electric valve. The air forces the piston forwards, which in turn moves the jaws of the gripper. Pneumatic systems are often used for grippers because gases compress and make the jaws "springy".

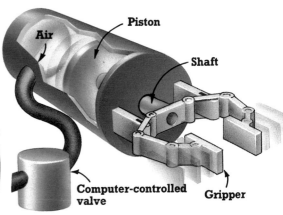

Piston

Air

Shaft

Computer-controlled valve

Gripper

How to make your own pneumatic gripper

You need: an empty detergent bottle, ice-cream or margarine tub, balloon, two pencils, thick card, tape, two fine nails or pins, sharp knife, scissors.

1. Carefully hammer nails through pencils half way along. Wiggle them around to make them loose. Cut two squares of card as shown.

2. Cut holes in front and lid of tub in the places shown here, with sharp knife.

3. Tape card to ends of pencils and push pencils through holes in the front of tub. Make sure nails are vertical and tape them to front of tub.

4. Remove cap of detergent bottle and put balloon on top. Push through hole in lid and between pencils. Squeeze bottle to inflate balloon and move "jaws".

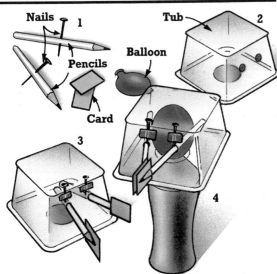

Nails

1

Pencils

Balloon

Card

3

Tub

2

4

25

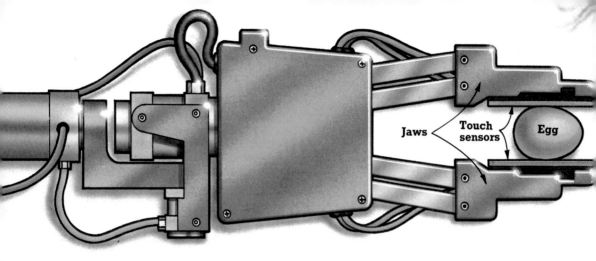

Jaws · Touch sensors · Egg

How robots hold things

Arm robots need a gripper, called an end effector, or another kind of tool fixed to their wrist to be able to do any work. There are many different kinds of grippers and tools and they are often specially designed to do a particular job. These pages explain what some of them are and how they work.

Hands

Some robots have grippers with jaws for grasping things. The picture above shows a gripper with two jaws holding an egg. These have to be able to hold an object without either crushing it or letting it slip. This is difficult to control, the pressure must be right otherwise the robot tends to fling the object out of its grasp when it moves quickly. The jaws in the picture have touch, or tactile, sensors on them which tell the computer how tightly they are gripping so that the correct pressure can be applied.

How robots change tools

A robot may need different tools to do a job and its computer can be programmed to make it change them automatically. The robot below holds tools with a bayonet fitting, like a light-bulb holder, attached to its wrist. The robot lowers the tool to be changed into a cradle, which holds it while the robot twists its wrist and pulls its arm away. The whole process is reversed to pick up another tool for the next stage in a job.

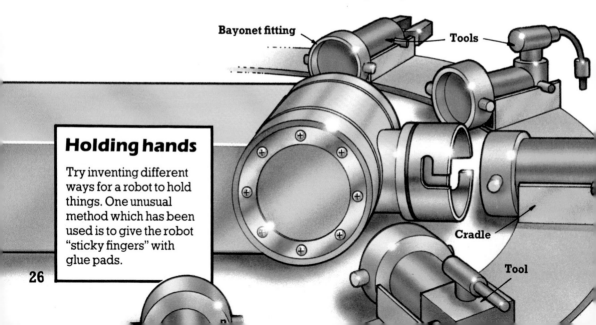

Bayonet fitting · Tools · Cradle · Tool

Holding hands

Try inventing different ways for a robot to hold things. One unusual method which has been used is to give the robot "sticky fingers" with glue pads.

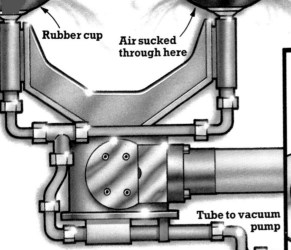

Rubber cup   Air sucked
              through here

Magnet grippers

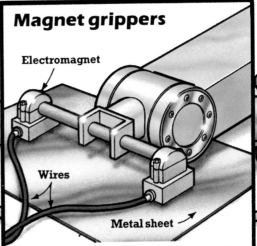

Electromagnet

Wires

Metal sheet

Tube to vacuum
pump

Vacuum grippers

Vacuum grippers, like the ones above, are
often used to pick up fragile objects – glass,
or paper sacks, for example. The grippers
are rubber cups and air is sucked through
them in much the same way as a vacuum
cleaner. This makes the object stick to the
gripper. The flow of air is controlled by the
computer and the weight the grippers are
able to lift depends on how powerful the
suction is.

Electric magnets, called electromagnets,
are sometimes used as grippers for
picking up metal objects. They are
connected to a supply of electricity and
become magnetic only when the power is
switched on by the robot's computer.
They lose their magnetism when the
power is switched off, and drop what they
were holding.

Holding tools

Many kinds of tools can be bolted directly
to the robot's wrist. The robot below, for
example, has a small electric grinder fixed
to its wrist to take rough edges off pieces of
metal. This method of holding a tool is used
where the robot has to do the same job over
and over again without the need for a tool
change.

Make your own electromagnet

You can find out how an electromagnet
works by making one with a large steel
nail, a length of plastic covered wire and a
battery.

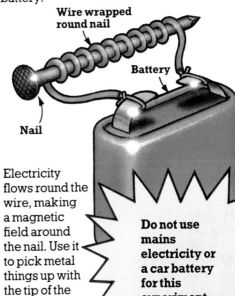

Wire wrapped
round nail

Battery

Nail

Electricity
flows round the
wire, making
a magnetic
field around
the nail. Use it
to pick metal
things up with
the tip of the
nail.

**Do not use
mains
electricity or
a car battery
for this
experiment.**

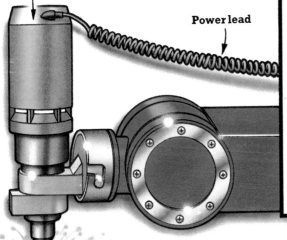

Electric grinder

Power lead

Computer control

Computers are programmed to control robots by sending them instructions, which take the form of electrical signals. Computers can also be programmed to react to information from the robot's sensors. These pages explain how an arm robot's computer instructs it to assemble things on a conveyor belt. It also shows how messages from a TV camera sensor make the computer interrupt its instructions to the robot when something goes wrong on the conveyor – when a sleeping cat comes along, for example. All robots are computer-controlled and what they can do depends upon the program used.

Computer messages

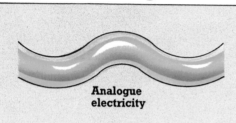

Analogue electricity

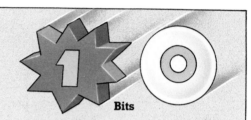

Bits

Most robot motors and sensors work with electricity which is a continuous wave – like the shape a skipping rope makes if two people hold the ends and wiggle the rope up and down. Information in this form is known as "analogue". Computers also use electrical signals but the information is in a different form.

Computers work using individual pulses of electricity called bits. There are two kinds – "no pulse" bits, written as O, which have a very low electric current, and "pulse" bits, written as 1, which have a stronger current. Information in this form is known as "digital" and the 1s and 0s make up a counting system called binary.

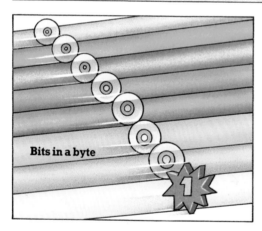

Bits in a byte

The computer is connected to the robot by eight or more wires called the bus. Many computers use groups of eight bits, called bytes, to represent pieces of information. Each byte is an eight-digit code made up of 0s and 1s. Bytes of information from the computer to the motors, and the sensors to the computer, take turns to go along the same bus in opposite directions. This picture shows the eight bits of a byte travelling parallel to each other along a bus.

The digital instructions sent out by the computer go to an interface which has an electronic switch for each of the robot's motors. This picture shows what happens at one switch. The "pulse" bit in the byte turns the switch for one motor either on or off. This allows analogue electricity to flow through the switch and along a wire to the motor. An interface is necessary because the motors do not work with digital pulses.

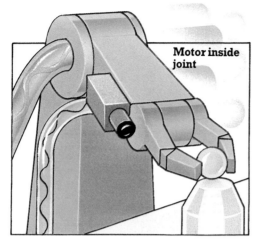

The analogue electricity powers a motor in one of the robot's joints – here it makes the arm go down to assemble two things on the conveyor.

The TV camera sensor on the side of the robot's arm sends pictures, in the form of analogue electricity, of the scene on the conveyor to the computer.

The analogue information from the TV camera is converted into digital information so that the computer can understand it. This is done by an interface, often called an analogue/digital converter.

The computer is programmed to analyse the information from the sensor and to react by switching the robot off if it cannot identify an object on the conveyor, like a sleeping cat.

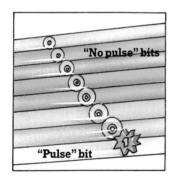

The computer sends out a byte with a "pulse" bit to the motor switching interface, just as it did to turn the motor on.

The "pulse" bit switches off the analogue electricity flowing through the switch. The motor stops because it has no power.

The whole process takes a split second, so the arm stops moving in time to avoid harming the cat.

Sensors

A robot's computer cannot know what is happening to the robot, or whether the robot has obeyed its instructions, unless it is equipped with sensors. There are two main types – those the robot uses to "touch" with, called contact sensors, and those used to "see" or "hear" with, called non-contact sensors.

Sensors work by sending an electric signal to the computer. The amount of electrical information the sensor sends out, called its output, depends on what the robot's environment does to it. A microphone "ear" would send a lot of information to the computer if someone standing next to the robot screamed, for example. Generally, the more that happens to the sensor, the greater its output.

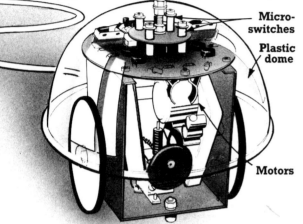

Micro-switches

Plastic dome

Motors

Hebot robot

The micro-robot on the left is equipped with one of the simplest kinds of contact sensor – a switch. There are four switches under the top of the plastic dome, which is loose. When the robot collides with something, the dome touches one of the switches, which sends a signal to the computer to reverse the robot's motors.

Hebot can be programmed to move around like the Turtle and can also be equipped with a pen to draw pictures. The pen can be lifted up and down under the control of the computer.

Touch sensors

Touch, or tactile, sensors tell the computer when, and by how much, the robot is touching something. These sensors are often used on grippers and on the bumpers, or fenders, of mobile robots. The computer needs feedback from these sensors so that it can control the

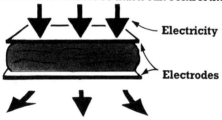

Electricity

Electrodes

robot not to crush whatever it touches. The picture above shows a tactile sensor made from a sandwich of a special foam rubber between two pieces of metal that conduct electricity, called electrodes. When nothing is touching the electrodes, the foam in the middle stops electricity running from one to the other. When the sandwich is squeezed, some of the electricity gets through, and this is converted by the computer into a measurement of pressure.

A light load on the sensor allows a small amount of current to pass from one electrode to another.

A heavy load squeezes the sensor more and lets a lot of electricity pass across the electrodes.

Another touch sensor

This tactile sensor works by using two optical fibres inside a cylinder. Optical fibres are thin tubes of glass used to transmit light. The cylinder has a flexible mirror at one end and two holes in the other. A lamp shines a beam of light down one fibre onto the mirror, which reflects it up the other fibre, and into a photoelectric cell. This detects the amount of reflected light. When the flexible mirror is pressed, less light is reflected into the photoelectric cell. The computer can convert the amount of reflected light into a measurement of pressure on the mirror.

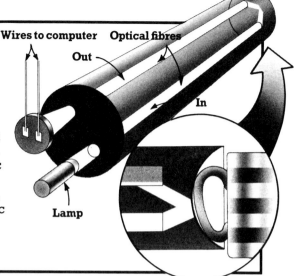

Wires to computer   Optical fibres

Out

In

Lamp

Vision

One of the most powerful non-contact sensors is vision. The picture on the right shows a kind of camera, called a solid state camera, connected to a computer. On the computer screen is the camera's view of a face.

The camera "views" an object with a grid of small, square, light-sensitive cells each of which corresponds to a square on the screen. Each cell is electrically charged. Light areas of an object viewed with the camera make the cells lose a lot of their charge, while dark areas only make them lose a little. The computer converts the charge on each cell into a square of light on the screen.

Solid-state cameras often have over 65,000 cells.

Camera

This robot has a camera fixed near its gripper to "look" for imperfect biscuits. Its computer analyses what the robot "sees" and instructs the robot to remove biscuits from the conveyor which are not up to standard.

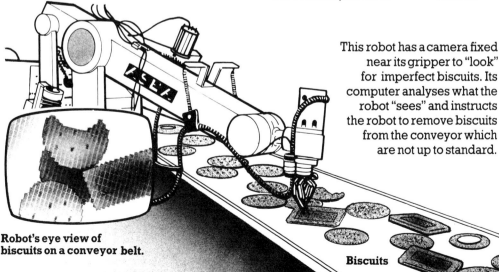

Robot's eye view of biscuits on a conveyor belt.

Biscuits

31

How a robot knows where it is

These pages explain how two different kinds of sensors are used to make measurements which tell robots where they are. Environmental sensors measure by how far a robot is from something else and positional sensors measure by how much part of a robot has moved.

How a mobile robot knows where it is

A mobile robot has to be equipped with environmental sensors which measure distances between the robot and other objects to find out where it is. One way of doing this is with an ultrasonic sensor which makes "time of flight" recordings. The picture below shows an ultrasonic sensor on an experimental tractor. It works by transmitting a "bleep" of sound and then receiving its echo which bounces back from surrounding objects. The robot's computer works out a distance from the time it takes for the echo to come back.

Puzzle

Sound travels about 330 metres (about 360 yards) in one second. It takes 1½ seconds for the sound to go from this robot's sensor to the tree and back again. How far away is the tree from the robot?

This "bleep" is too high for humans to hear.

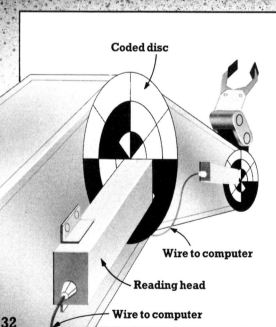

Coded disc

Wire to computer

Reading head

Wire to computer

How robots measure bends

An arm robot's computer needs to know that the robot has carried out its instructions by finding out the position of the arm. This robot has sensors on each of its joints which measure how much it has bent its arm and wrist. The sensor, called an optical position encoder, sends a digital message to the computer which it converts into an angular measurement. The sensor has two parts – a flat disc with marks on it, and a reading head which "reads" the marks. Each segment on the disc represents a number in binary code. The disc is attached to a part of the robot which moves, and the reading head is fixed to a part which stays still. As the robot bends its arm a different number is "read" from the disc and this number is sent to the computer.

How an arm robot knows where it is

Sensors can be fixed to an arm robot to give its computer arm-movement measurements. Some sensors are used to make straight-line measurement, and others to measure angles.

One way of making measurements is with a sensor called an electrical potentiometer. This works like a dimmer switch by varying the amount of electricity passing along a wire. The amount of electricity getting through the wire can be converted into a measurement by connecting it via an interface to the computer.

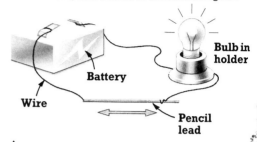

Bulb in holder

Battery

Wire

Pencil lead

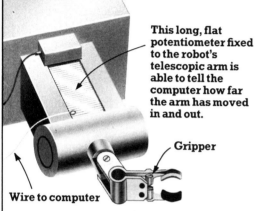

This long, flat potentiometer fixed to the robot's telescopic arm is able to tell the computer how far the arm has moved in and out.

Gripper

Wire to computer

▲
You can test the principle of a potentiometer using a battery, wire, a lamp and a pencil. Carefully split the pencil down the middle, take the lead out and connect everything together as shown. By sliding one end of the wire up and down the pencil lead, you can vary the brightness of the bulb. This happens because the lead resists the flow of electricity.

A special kind of wire which does the same thing as the pencil lead in the experiment shown above, is used in a potentiometer.

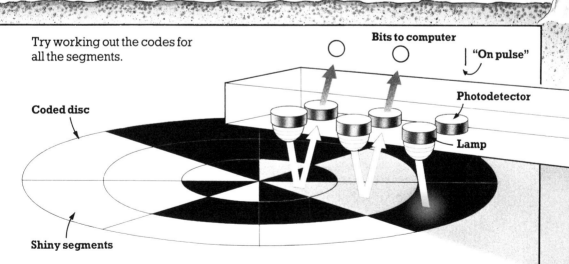

Try working out the codes for all the segments.

Bits to computer

"On pulse"

Photodetector

Coded disc

Lamp

Shiny segments

The reading head has three pairs of photodetectors and lamps. Shiny white parts on the disc reflect the light into the photodetectors. Shiny parts are registered as 0's, and black parts as 1's. Segment 1, for example, would be "read" as 001. This is called a 3-bit code because the sensor sends three bits to the computer.

Cybernetics

Cybernetics is the science of control and communication in both machines and living organisms. The word comes from a Greek word meaning "steersman". It is particularly concerned with things which are self-controlling, or adaptive. An adaptive system alters its behaviour because of changes in its environment. For example, George, an automatic pilot used in aeroplanes, alters the course of the plane as a result of changes in the wind speed.

Artificial intelligence

A closely related field to cybernetics is artificial intelligence (AI) which is about making machines do intelligent things. Machines have to be able to "think" to do something intelligent, but experts disagree about what this means. There are some who believe that a machine which "learns" from past experience, or responds to things happening to it, like George, can be called a "thinking" machine. Others argue that for machines to think they must have feelings and want to do things. This would mean that a "thinking" robot, for instance, would have to want to pack boxes because it enjoyed its work.

Clever machines

Computers are the cleverest machines available because by ingenious programming they can be made to simulate, or mimic, intelligent human activity, such as the processing of visual information and speech. Computers can then be used to control other machines, like robots, to make them behave "intelligently".

"Intelligent" computer

There are two basic ways of programming a computer. Algorithmic programs – often used for robots – work by considering all the possible alternatives in a situation. Heuristic programs are "cleverer" because they take short-cuts to decisions by remembering from past experience the best way to solve a problem. A chess-playing robot computer could work out the best moves by being given the rules of the game, for example. AI programs are often heuristic.

Speech recognition

Computer programs are being developed to give robots the ability to recognize spoken commands, using a microphone as an electronic "ear". The average adult knows thousands of words, so it would take a computer with a massive memory to understand even a tiny fraction of them. The computer also has to take into account the different ways that people speak. It is much simpler to program the computer to recognize only a short list of words, spoken by one person, which are needed for the robot's job.

How computers recognize words

1. Each word makes wave-like patterns of sound that are converted by a microphone into electricity. The waves vary according to the different sounds in a word.

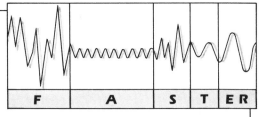

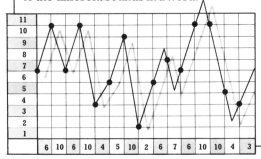

| | | | | | | | | | | | | | | |
|6|10|6|10|4|5|10|2|6|7|6|10|10|4|3|

2. The height of the wave, which is an electrical voltage, is measured many times a second. These measurements are recorded as a sequence of numbers, and then turned into a digital code of "pulse" and "no pulse" bits which the computer can then use to identify the word. The picture on the left shows how a word like "faster" would look to the computer.

Vision

Robots are increasingly being equipped with machine vision which allows them to "see" and behave "intelligently". The intelligent part of this is not the TV camera eye, the computer brain, or the robot, but the computer program. This analyses and interprets what the "eye" sees – something which is extremely complicated. Humans are very selective in what they actually see, and this is difficult to simulate with a computer. For example, if you look carefully at this picture you will be able to choose whether you see either a vase or two faces. Machine vision could not do this.

How robots recognize things

A machine vision system can be programmed to recognize one or more objects. This shows how an object in a pile can be recognized, so that a vision system can tell a robot how to pick it up correctly for packing in a box.

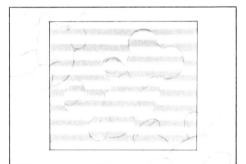

The system focuses on one part of the pile and projects stripes of light over it to judge how far away it is. This information is sent to the robot's computer.

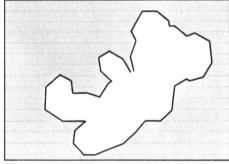

The computer can work out the outline of one bear from the breaks in the stripes of light. It is programmed only to identify the outline of bears and will not recognize anything else.

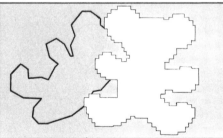

By comparing this outline with views of the bear stored in its memory, the computer can work out the position of the bear in the pile. This information is then sent to the robot in the form of instructions to its motors.

The computer controls the robot to pick the bear up without damaging it or any of the other bears. It then turns the bear the right way round for packing. This sequence is repeated for all the bears.

Latest developments

Robotics is a fast-moving and exciting subject with many research projects going on around the world. More and more arm robots are being used in factories along with other automatic machines. Robots are also being made more "intelligent" by using more and better sensors together with clever computer programs for their control. This means mobile and other kinds of robots may soon become more familiar – perhaps in the home as robot "servants" and in factories too. Robots are also becoming cheaper – a micro-robot costs about the same as some home computers. Some of the latest developments are shown here.

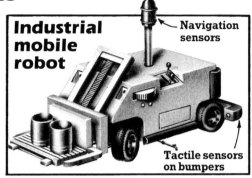

Industrial mobile robot

Navigation sensors

Tactile sensors on bumpers

This is a driverless forklift truck which will be used in an automated warehouse or factory. It has an on-board computer and power supply and uses sensors to navigate.

Nuclear reactor robot

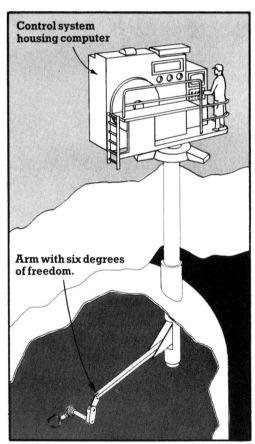

Control system housing computer

Arm with six degrees of freedom.

This arm robot is designed to be used in the core of a nuclear reactor. The arm is suspended from a long hollow chain. Control cables for the arm pass through the chain.

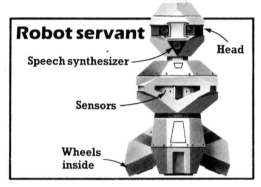

Robot servant

Speech synthesizer

Head

Sensors

Wheels inside

This robot can be programmed to do things like serve drinks at a party and speak to guests with its synthesized voice. Others are being made which do housework.

Robot metro

A completely automatic robot train has been built in Lille, France. The trains are computer-controlled to switch between tracks and are programmed to stop automatically at stations.

Robot computer assistant

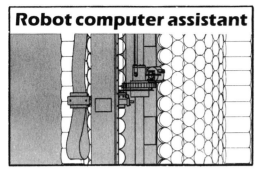

This arm robot goes up and down in a honeycomb storage cell to find special cartridges containing computer data. It delivers the cartridges to the computer and replaces them after use.

Two-armed robot

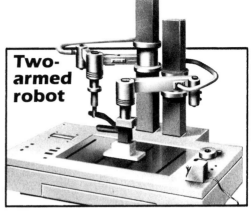

Yes-Man is designed to work alongside humans on a production line. Its arms allow it to do complex assembly work – it can even do two things at once. The base contains control microcomputers.

Walking robot

A four-legged walking robot which can climb stairs has been built by Japanese scientists. Other researchers are trying out six- and eight-legged designs which walk like insects.

Modular arm robot

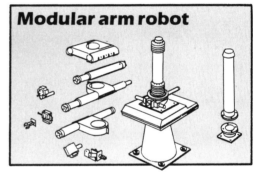

Some arm robots are being made in modules – small units, such as arm, wrist, base and so on – that can be combined in different ways to make a robot suitable for a particular job.

Robot cleaning machine

Twin scrubbing brushes

A free-roving, industrial floor-scrubbing robot is being developed. As well as navigation sensors, it will probably have a sensor to detect when the water becomes dirty.

Android

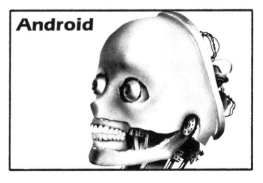

Androids – robots which look and act like humans – are being made, mostly for exhibitions and shop displays. This one is powered by electric motors and hydraulic pistons.

Build your own micro-robot

The next nine pages show how to build a computer-controlled micro-robot. You need a computer with a parallel input/output port to control the robot (see page 46 to find out if your computer is suitable). The project gives step by step instructions on how to make an electronic interface circuit to connect the robot to a computer, with hints on soldering and components. A computer program is included which will control the robot to move like a Turtle or a Bigtrak (see page 16).

The robot is made using a flat base-board with two motors, two gearboxes and two wheels mounted on it. It also has a small wheel at the back which stops the robot from tipping up. The wheels are driven by separate motors, via gearboxes which reduce the speed of the motors. The computer steers the robot by controlling the direction of the two motors. See page 6 to see how this works. You could put a

cardboard body over the base, and this can be any shape you like. Page 1 shows a picture of a home-made robot mouse, for example.

The robot can be made to turn left and right, and to go forwards and backwards. The computer program in this project lets you give the robot a sequence of instructions to move it in any direction you like. By attaching a pen to the robot with tape, you can make it draw pictures.

The electronic part of this project is not easy to build. A single faulty component or a tiny mistake could prevent the robot from working. The robot itself can be built using parts from a construction kit, such as Fischertechnic. The robot below is made in this way, but other methods are suggested. The project may be quite expensive, depending on whether you already have a construction kit. It is a good idea to work out the cost of all the components before starting.

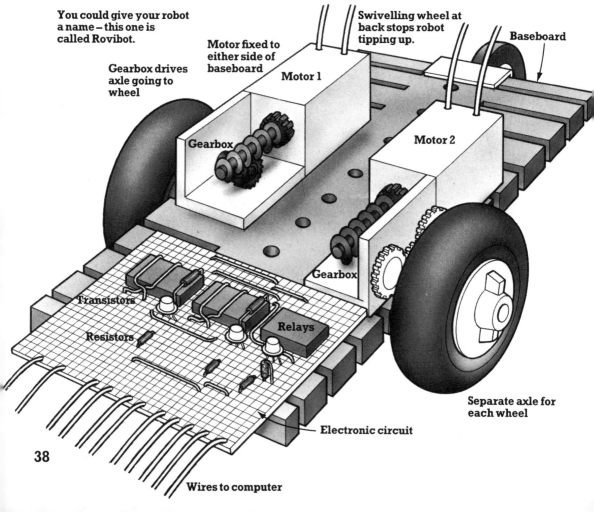

You could give your robot a name – this one is called Rovibot.

Swivelling wheel at back stops robot tipping up.

Baseboard

Motor fixed to either side of baseboard

Gearbox drives axle going to wheel

Motor 1

Gearbox

Motor 2

Transistors

Resistors

Relays

Gearbox

Separate axle for each wheel

Electronic circuit

Wires to computer

Components for the project

You can buy components in an electronic components shop, or you can buy a mail order kit of either the electronic or the mechanical parts for the project from a supplier shown on page 46. Ask in your local TV repair shop if you are not sure where the nearest component shop is. It is a good idea to take this book with you.

Parts for robot

2 × motors with a voltage range between 3V-12V (construction kit motors, like Fischertechnic, are ideal but you could also use motors from an old battery-powered toy car, or buy motors from a model shop).
2 × gearboxes which match the motors (i.e. if you use Fischertechnic motors you will need the same make of gearbox).
2 × wheels and axles (make sure they will fit the gearbox).
1 × small swivelling wheel.
Baseplate (use a piece of plywood about 100mm × 200mm × 10mm if you do not have a construction kit).

Parts for electronic circuit

2 × double-pole changeover relays, coil voltage 6V d.c., coil resistance greater than 150Ω (250Ω is best), suitable for 0.1 pitch Veroboard
1 × single-pole relay with the same specifications as above (see notes on relays on page 40).
3 × transistors 2N222A, BC107 or BC108 or any NPN transistor with a current gain (HFE factor) greater than 100.
3 × 2.2K Ω resistors
3 × diodes IN4001, IN4002 or IN4003. (Do not use Zener diodes).
Veroboard with copper strips, size 0.1 inch pitch, 30 tracks × 26 holes or Prototype board, which you do not need to solder.

Other things you need

Soldering iron, cored solder, wire cutters, wire strippers, thin-nosed pliers, 22m of thin electric wire ("bell wire" or stranded wire is best), electrical tape, dress-maker's pins, pencils, tracing paper, paper glue, damp sponge,
4.5mm twist drill.

Power supply

Use battery or transformer power supply. DO NOT use car batteries or mains electricity as this is very dangerous. The power supply must match the voltage of the motors you use (i.e. 6V motors need a 6V battery or a 6V transformer).

About the electronic components

The electronic components you buy may not look the same as the ones drawn in this book. Some components MUST be connected a certain way round. Many components have marks or tags on them to identify particular legs, others come with diagrams. Some diagrams are labelled "pin view", which means you have to look at the component upside down, with its pins facing you, to identify them.

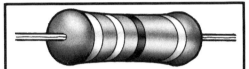

Resistors: These are used to reduce the amount of current in a circuit. It does not matter which way round they go. Colour coded stripes on the resistor show how many ohms (written Ω or KΩ for 1,000 ohms) it is.

Stripe this end

Diodes: These allow current to flow in one direction only – a bit like a one way street for electricity. Diodes only work one way round, so they have a stripe at one end to identify which way they should go.

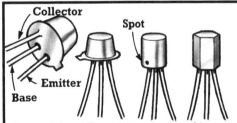

Collector   Spot   Emitter   Base

Transistors: Transistors are used in this project as switches to turn current on and off. They have three legs, a "collector", an "emitter" and a "base", and they must be connected up the right way. The centre leg is usually the base and the emitter is usually next to a tag, or other mark, on the case of the transistor. The transistors in this project are switched on and off by the computer.

Relays

Relays are electronic switches, activated by an electromagnet. Two types are used in the project – a single-pole relay with one switch inside, and a double-pole relay with two switches inside. When the electromagnet is off, the switch, or switches, stay in one position. When the electromagnet is on, the magnetic field pushes the switch into another position. The electromagnets in the relays used in this project are switched on and off by transistors.

It is not usually possible to see how to connect a relay by looking at the case or the pins underneath. Ask for a circuit diagram showing the inside. Pin connections vary according to manufacturer and type of relay. These pictures show the type used in this project and their circuits. Look carefully at the circuit diagram for your relays and substitute the pin numbers with those used below. Unless you do this, your relay may not match the pin numbers used in the instructions for the circuit.

Sub-miniature change-over relays

Single pole

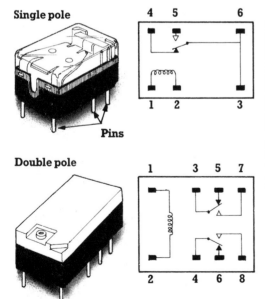

Pins

Double pole

If the diagram with your relay says "pin view", make sure you identify the pins with them facing you.

Check point

The instructions on this project are given for sub-miniature relays with the pins in the same positions as those shown on the guide below, as well as the same circuit inside. Check that the pins on your relay are in the right positions by putting them over the guides to see if they line up.

Single pole		Double pole			
• •	•	•	•	•	•
• •	•	•	•	•	•

If your relay does not line up with these guides, this is what to do:
Either: Turn the relay on its back and solder a piece of tinned wire (see hints on soldering on page 41) about 75mm long to each leg. Look carefully at the circuit diagram for your relay and substitute its pin numbers for those shown in the project. You can then solder the wires into the Veroboard instead of the pins.
Or: Look at the circuit diagram at the end of the project and work out a new circuit to suit the layout of the pins on your relay by looking at the diagram on page 46.

Veroboard

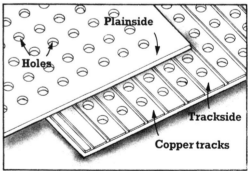

A special board, called Veroboard, is used to connect the electronic components together. This has rows of holes in it, with copper strips on the back linking them together. You push the legs of the components through the holes and solder them to this copper track. Electric current can then flow along the track between the components. Make sure the components do not touch each other on the Veroboard, especially the transistors.

Hints on soldering

Soldering is a way of joining two pieces of metal together, using another metal called solder, melted with a soldering iron. The picture on the left shows the things you need. Make sure the soldering iron is kept propped up when you are not using it so as not to burn anything.

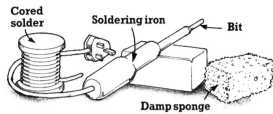

Cored solder • Soldering iron • Bit

Damp sponge

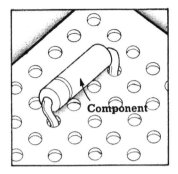

Component

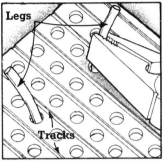

Legs
Tracks

1. Push the legs of the components through the holes on the plainside of the Veroboard.

2. Turn the Veroboard over, and bend the legs out slightly, using the pliers.

3. Wipe the bit on the damp sponge to remove old solder.

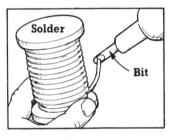

Solder
Bit

Remove any solder between tracks by running the hot bit along in the groove.

4. Touch the bit with solder so that a drop clings to it to "wet" it. This is called "wetting" the bit.

5. Carefully touch the bit on one side of the leg where it touches the track, while at the same time touching the solder on the other side of the leg. Hold them there for about one second until a small blob of solder flows around the leg. Let the joint cool for a few seconds while the solder hardens.

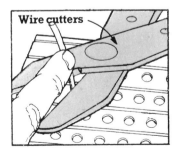

Wire cutters

Bit

Do not forget to unplug the soldering iron when finished.

Pliers

6. Trim the legs close to the solder with wire cutters. Hold the board away from your face and put your finger on the leg to stop it flying up in the air.

Wires: The ends of pieces of wire should be covered with solder. This is called "tinning". Stroke the wire quickly with the bit and the solder at the same time until the wire is lightly coated with solder. Tinning is done to get a good connection when soldering. It also holds together the strands of stranded wire to stop them unwinding. Tin the area of wire you have stripped.

41

Motor control circuit

The motor control circuit enables the computer to switch both motors on or off, or to control each motor independently to go forwards or reverse. The instructions have to be followed very accurately for the circuit to work.

Drill bit

Guide ▽

◄── **Tracks run horizontally** ──►

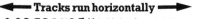

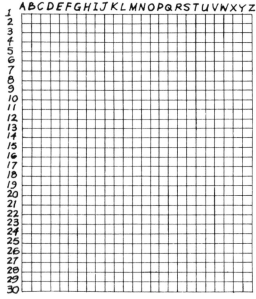

1. Photocopy or trace the positioning guide above, then cut it out.

2. Put a dab of glue under each corner of the guide.

3. Place the guide on the plainside of the Veroboard by pushing a pin through holes A1 and Z30 to help line up the guide with the holes and tracks of the Veroboard.

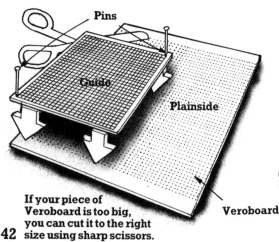

Pins

Guide

Plainside

Veroboard

If your piece of Veroboard is too big, you can cut it to the right size using sharp scissors.

4. Cut the track with a 4.5mm drill bit at holes H2, H3, H7, H10, H13, H15, H17, H20, H23, H25, H27, Q4, Q9, Q19. Hold the bit in your fingers and turn it to remove all the copper from the track.

5. Place a single-pole relay on the Veroboard with the pins in these holes:
Pin 1: J2
Pin 2: J3
Pin 3: J7
Pin 4: G2
Pin 5: G3
Pin 6: G7
Solder each pin, taking care not to join the tracks with solder.

6. Place a double-pole relay on the Veroboard in these holes and solder each pin.
Pin 1: G10
Pin 2: J10
Pin 3: G13
Pin 4: J13
Pin 5: G15
Pin 6: J15
Pin 7: G17
Pin 8: J17

7.
Place a double-pole relay on the Veroboard in these holes and solder each pin:
Pin 1: G20
Pin 2: J20
Pin 3: G23
Pin 4: J23
Pin 5: G25
Pin 6: J25
Pin 7: G27
Pin 8: J27

Relay with wires soldered to pins

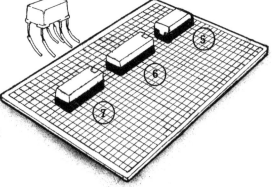

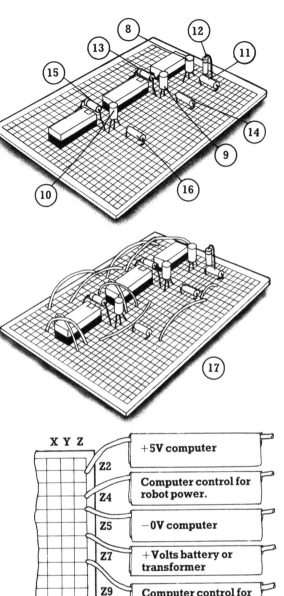

8. Solder the collector leg of a transistor in hole M3, the base leg in hole M4 and the emitter leg in hole M5.

9. Solder the emitter leg of a transistor in hole M8, the base leg in hole M9 and the collector leg in hole M10.

10. Solder the emitter leg of a transistor in hole M18, the base leg in hole M19 and the collector leg in hole M20.

11. Solder a resistor with one leg in hole P4 and the other in hole S4.

12. Solder a diode with the leg nearest the striped end in hole P2 and the other in hole P3.

13. Solder a diode with the leg nearest the striped end in hole E10 and other leg in hole L10.

14. Solder a resistor with one leg in hole P9 and the other in hole S9.

15. Solder a diode with the leg nearest the striped end in hole E20, and the other in hole L20.

16. Solder a resistor with one leg in hole P19 and the other in hole S19.

17. Cut 11 lengths of wire about 100mm long. Strip each end about 10mm. Tin both ends of each wire. Then loop one wire between each of the following pairs of holes and solder them into place as you go:

L2 and D10, C2 and C13, B13 and B23, C10 and C20, E15 and L17, E17 and L15, E25 and L27, E27 and L25, T5 and T8, U8 and U18, M13 and M23.

18. Cut 7 lengths of wire about 3m long. Strip each end about 10mm and tin one end. Label each wire with a piece of tape as shown in the white labels on the left. Label the other end of each wire with the same label. Solder the tinned end of each wire into the holes shown on the left. Label each wire as you go, otherwise you might get muddled up.

19. Cut 4 lengths of wire about 250mm long. Strip each end about 10mm and tin one end. Label each wire with a piece of tape, as shown in the shaded labels on the left. Solder the tinned end of each wire into the holes shown next to these labels.

43

How to connect circuit to computer, motors and power

Computer: Connect circuit to computer's parallel input/output port by the wires soldered at step 18 in the instructions. You will probably need to buy an edge connector to plug into the port. You can get these from component or computer shops. Use computer's handbook to identify pins in the port, and connect up to wires shown in white spaces in the chart below.

Motors: Connect wires soldered at step 19 in the instructions to the terminals of the motors, as shown in shaded spaces in the chart.

Power supply: Connect the last two wires labelled in the chart to the + and − terminals of your battery or transformer.

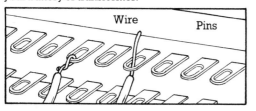

Wire   Pins

Push the wires through the holes in the pins on the edge connector and twist them round, making sure that the wires do not touch each other. Do not solder.

Wire label	Connection
+5V computer	5 volt pin of user port
−0V computer	0 volt pin of user port
Computer control for motor 1	PB2 pin of user port
Computer control for motor 2	PB0 pin of user port
Computer control for robot power	PB1 pin of user port
Motor 1A	Right hand terminal of motor 1
Motor 1B	Left hand terminal of motor 1
Motor 2A	Right hand terminal of motor 2
Motor 2B	Left hand terminal of motor 2
+volts battery/ transformer	+ volt terminal of battery or transformer
−volts battery/ transformer	−volt terminal of battery or transformer

Computer program

The computer program opposite allows the robot to go forwards, backwards, left or right, so many units at a time. You will have to experiment to find out how far each unit is because it depends on the number you set in line 650 of the program. The larger the number, the further each unit will make the robot go. This menu will appear on the screen once you have entered the program.

1. Tell me what to do
2. Go
3. Clear memory

If you press 1, then RETURN, you can give the robot any of these instructions:**Forwards:** Press F, then RETURN, then a number, then RETURN. Pressing F RETURN 6 RETURN will make the robot ready to go forward 6 units, for example.
Backwards: Press B, then RETURN, then a number, then RETURN.
Left: Press L, then RETURN, then a number, then RETURN.
Right: Press R, then RETURN, then a number, then RETURN.
Stop: Press S and program will go back to the menu.

To make the robot carry out your instructions, press 2, then RETURN. You can give it a sequence of instructions, for example, forwards 5, left 3, forwards 6, back 2 and so on. After pressing 2 to make the robot go, the instructions will be displayed on the screen as the robot moves. To give the robot new instructions, press 3, then RETURN.

Adjusting the program

Before entering the program in your computer you will need to do some tests to see what numbers to enter in lines 580, 590, 600, 610, 690 and 740.
1. Connect up as shown on this page.
2. Type this program into your computer.

```
10 ?&FE62=7
20 LET OL=&FE60
30 INPUT P
40 ?OL=P
```

3. Type the numbers 0 to 7 one at a time. Look to see which direction the robot's motors run in response to each number. Write the number which makes the motors run in the correct direction into the program, at the lines shown below.
580 – Both motors forward
590 – Both motors backwards
600 – Motor 1 forwards, motor 2 backwards.
610 – Motor 1 backwards, motor 2 forwards
690 and 740 – both motors off

THIS PROGRAM IS WRITTEN FOR THE BBC COMPUTER. THE SYMBOLS IN THE LEFTHAND COLUMN SHOW WHERE CHANGES HAVE TO BE MADE SO THAT IT WILL RUN ON OTHER COMPUTERS. THESE CHANGES ARE LISTED ON THE NEXT PAGE.

```
◇▲ 10 ?&FE62=7
◇▲ 20 LET OL=&FE60
   30 DIM D(20)
   40 DIM M$(20)
   50 GOSUB 550
▲  60 CLS
   70 PRINT "ROBOT CONTROL"
   80 PRINT
   90 PRINT "1. TELL ME WHAT TO DO"
  100 PRINT "2. GO"
  110 PRINT "3. CLEAR MEMORY"
  120 PRINT
  130 PRINT "TYPE NUMBER"
  140 INPUT C
  150 IF C<1 OR C>3 THEN GOTO 130
◇ 160 ON C GOSUB 180,440,550
  170 GOTO 60
  180 LET PC=PS
▲ 190 CLS
  200 IF PC=20 THEN GOTO 390
  210 PRINT
  220 PRINT "INPUT STEP ";PC
  230 PRINT "DIRECTION THEN TIME"
  240 INPUT M$(PC)
  250 IF M$(PC)="S" THEN GOTO 410
  260 INPUT D(PC)
  270 LET P=999
  280 GOSUB 580
  290 IF P<>999 AND D(PC)>0 THEN GOTO 320
  300 PRINT "WRONG COMMAND"
  310 GOTO 220
  320 GOSUB 630
▲ 330 CLS
  340 FOR I=1 TO PC
  350 PRINT "STEP ";I;": ";M$(I);" ";D(I)
  360 NEXT I
  370 LET PC=PC+1
  380 GOTO 200
  390 PRINT "NO MORE STEPS"
  400 LET M$(PC)="S"
  410 LET PS=PC
  420 GOSUB 710
  430 RETURN
▲ 440 CLS
  450 LET PC=1
  460 PRINT "STEP ";PC;": ";M$(PC);" ";D(PC)
  470 IF M$(PC)="S" THEN GOTO 520
  480 GOSUB 580
  490 GOSUB 630
  500 LET PC=PC+1
  510 GOTO 460
  520 PRINT "END OF INSTRUCTIONS"
  530 GOSUB 710
  540 RETURN
  550 LET M$(1)="S"
  560 LET PS=1
  570 RETURN
  580 IF M$(PC)="F" THEN LET P=1
  590 IF M$(PC)="B" THEN LET P=2
  600 IF M$(PC)="R" THEN LET P=0
  610 IF M$(PC)="L" THEN LET P=3
  620 RETURN
◇▲ 630 ?OL=P
  640 FOR J=1 TO D(PC)
  650 FOR L=1 TO 100
  660 NEXT L
◇▲ 670 IF INKEY$(0)="S" THEN GOTO 740
  680 NEXT J
◇▲ 690 ?OL=4
  700 RETURN
  710 PRINT "PRESS RETURN FOR MENU"
  720 INPUT Z$
  730 RETURN
◇▲ 740 ?OL=4
  750 STOP
```

Arranges output and makes space for instructions to the robot.

Prints menu on the screen.

Goes to part of the program that organizes instructions to the robot.

Lets you give the robot instructions. If you run out of memory for instructions, the program returns you to STOP.

Analyses the instructions and carries them out as long as they are valid commands.

Lists instructions given so far on the screen.

Goes back for the next instruction to the robot.

If the last instruction is stop, program goes back to the menu.

Carries out instructions to the robot after pressing 2.

If you press 3, the program clears the last set of instructions to the robot.

Looks at the instruction and decides what number to output.

Makes robot move according to the instructions given to it.

Waits for RETURN to be pressed.

Stops everything if you press S.

Input instructions

Instructions

Go

Clear

Analyse

Move

Wait

45

Changes for other computers

▲ VIC 20   ◇ ZX81 (Timex 1000)

◇ 10 DELETE

▲ 10 POKE 37138,7

◇ 20 LET OL=NUMBER OF MEMORY LOCATION FOR OUTPUT

▲ 20 LET OL=37136

▲ 60,190,330,440 PRINT CHR$(147)

◇ 160 GOSUB 180*(C=1)+440*(C=2)+550*(C=3)

◇▲ 630 POKE OL,P

◇ 670 IF INKEY$="S" THEN GOTO 740

▲ 670 GET A$:IF A$="S" THEN GOTO 740

◇▲ 690,740 POKE OL,P

Wiring diagram

Look at the diagram and work out a new circuit if your relays do not suit the instructions.

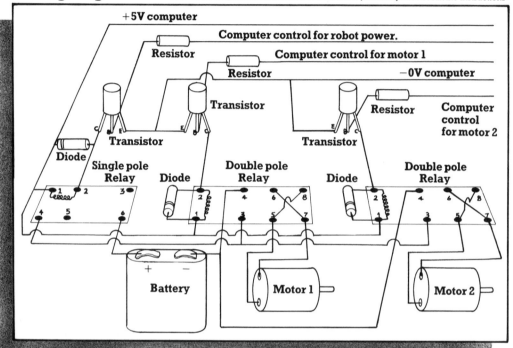

Computers you can use with this project

BBC Model B
Commodore VIC 20
*Sinclair ZX81 (Timex 1000)
*Sinclair Spectrum (Timex 2000)
*You need a special interface for these computers. You can buy one by mail order from this company.
Colne Robotics Co Ltd, Beaufort Road, off Richmond Road, East Twickenham, Middlesex, TW1 2PQ, England.
Or you can get a kit (not easy to assemble) from:
Powertran Cybernetics, Portway Industrial Estate, Andover, Hampshire, England.
Also look in computer magazines for advertisements

Complete kit of parts for robot and circuit

You can buy a kit of parts by mail order from:
Bluepond Electronics, Alpha Road, Crawley, Sussex, England.

Send a stamped addressed envelope to these companies for details of their products.

What to do if the robot does not work

Carefully check that you have all the components in the correct place and resolder any which look loose. Make sure that all the wires are connected properly and that they don't touch each other. Check you have good batteries and the cut-out switch if you are using a transformer. If the robot still does not work, get someone else to look at it, as it is easy to miss something. Make sure that your motors work by connecting them directly to a suitable battery. If you still cannot get the circuit to work pack it carefully (with enough stamps for return postage) and send to:

Electronics Advisor,
Usborne Publishing,
20 Garrick Street,
London, WC2E 9BJ.

Robot words

Android: A kind of robot made to look human.

Artificial Intelligence: The study of making machines do "intelligent things". Experts disagree on a precise definition of what counts as intelligence or intelligent behaviour.

Degrees of freedom: A technical term used to describe the different directions an arm robot can move. Usually, the more joints a robot has in its arm, the more degrees of freedom it has.

Feedback: Information about the robot or its surroundings that a computer gets from sensors on the robot.

Gears: These reduce or increase the speed of a motor. They are used between a motor and the part of a robot that it drives.

Gripper: The mechanism fixed to an arm robot's wrist to hold things. Sometimes called an end effector.

Hydraulic system: A device using a special oil in pipes and cylinders to drive mechanical parts of a robot. Often used on arm robots.

Interface: Used between a robot and its computer to convert electrical signals from the computer into instructions from the robot and vice versa.

Lead-through programming: A way of teaching a robot by guiding it through the movements needed to do a job.

LOGO: A computer language often used to program robots which draw, like the Turtle.

Machine vision: A computer-controlled device used to give robots a primitive kind of sight.

Navigation: How a computer uses information from a mobile robot's sensors to get the robot from one place to another without bumping into anything.

Odometer: A sensor that measures the distance travelled by a wheeled vehicle.

Photoelectric cell: An electronic device which detects light. These are often used as part of a sensor on robots.

Pitch: The name for the up and down movement in a robot's wrist, which is like the movement made when using a lever.

Pneumatic system: A device powered by air or another gas to operate a mechanical part of a robot – often the gripper.

Port: The socket on a computer where interfaces and other kinds of electronic equipment are plugged in.

Program: A sequence of instructions given to a computer that controls everything the robot does.

Robot: A computer-controlled machine which can be programmed to do different kinds of things. Experts do not agree on an exact definition of a robot.

Roll: A name for the movement in a robot's wrist which goes from side to side like rocking a boat.

Sensor: A device which gives a robot's computer information either about the robot or its surroundings.

Sonar sensor: Often used for navigation, these sensors emit a sound and then "listen" for an echo to bounce back from obstacles. Distances are calculated by the time taken for the sound to return.

Speech synthesizer: An electronic device, often a chip, which can be programmed to produce words and sentences through a loudspeaker. Each word is broken down into small units of sound, which are then reproduced digitally.

Transformer: An electronic device which converts mains electricity into low voltage suitable for powering things like micro-robots and train sets.

Turtle: A wheeled micro-robot programmed to move about and draw, using a computer language called LOGO.

Working envelope: The area of space that an arm robot is able to reach.

Yaw: The name given to the left and right movement in a robot's wrist – similar to the movement made steering a bicycle.

Index

algorithmic programs, 34
analogue, 28, 29
analogue/digital converter, 29
androids, 37
Armdroid, 17
arm robots, 8, 12, 13, 14, 24, 32, 33, 36, 37
arms, 8, 10, 11, 32, 33, 37
artificial intelligence, 34

batteries, 13, 16, 39, 43, 44, 46
BBC Buggy, 17
BBC (Microcomputer), 46
Bigtrack, 6, 7, 38
binary, 28, 32, 33
bits, 28, 29, 33
bus, 28
bytes, 28, 29

cable-following robot, 6
Capek, Karel, 3
circuit diagram, 40
components, 12, 17, 38, 39, 40, 41
computer program, 4, 10, 20, 38, 44, 45
control lines, 16
control signals, 13
cruise missiles, 15
cybernetics, 34
cylindrical robots, 23

degrees of freedom, 8, 9, 36
digital information, 28, 29, 32, 34
diodes, 39, 43, 46
direct current (d.c.) motors, 24
double-pole changeover relays, 39, 40, 42, 46

elbow, 8, 14
electric motors, 8, 14, 24, 37
    d.c., 24
electromagnet, 24, 27, 40
end effector, 26
experimental robots, 12, 32

feedback, 4, 11, 30
Fischertechnic, 17, 38, 39
fork-lift truck, 19, 36

gearboxes, 38, 39
gears, 9, 24
George, 34
grippers, 9, 25, 31
    magnetic, 27
    vacuum, 27

hands, 12, 26
Hero 1, 13
heuristic programs, 34
home computer, 16
hydraulic system, 24, 25

infra-red transceiver, 17
"intelligence", 10, 34, 35, 36
interfaces, 8, 9, 16, 21, 28, 38
    analogue/digital 29

jaws, 25, 26
jointed-arm robots, 22
joints, 8, 9, 14, 29

lead-through programming, 20
legs, 10, 11
light detector, 6
light-sensitive cells, 31
Logo, 16, 17, 20

machine vision, 35
manipulative robot, 4
microchip, 3, 12
microphone, 3, 12, 34
micro-robots, 3, 5, 6, 16, 20, 36, 38
mobile robots, 6, 13, 24, 30, 32, 36
modular arm robot, 37
motor control circuit, 42
motors, 10, 16, 29, 35, 38, 39, 44, 46
Mr Bill, 7

navigation, 10, 11, 36, 37
"no pulse" bits, 28, 29, 34
nuclear-reactor robot, 36

odometers, 7
optical fibres, 31
optical position encoder, 32

parallel input/output port, 38, 44
photodetector, 33
photoelectric cell, 16, 31
pitch, 9
pneumatic system, 24, 25
polar robots, 22
port, 16
potentiometer, 33
Prototype board, 39
"pulse" bits, 28, 33, 34

relays, 39, 40, 42, 46
Remote Manipulator System (RMS), 14
resistors, 39, 43, 46
robota, 3
robot computer assistant, 37
robot mouse, 38
robot servant, 10, 36
robot sheep-shearer, 12
robot train, 36
roll, 9

satellites, 14, 15
sensors, 3, 5, 14, 16, 20, 28, 30, 37
    contact, 30
    environmental, 32
    non-contact, 30

positional, 32
    tactile, 26, 30, 36
shaft, 9, 15, 24, 25
shoulder, 8, 14
Sinclair (Timex) computers, 46
single-pole changeover relays, 39, 40, 42, 46
solder, 39, 41
soldering, 38, 40, 41, 42, 43
soldering iron, 39, 41
solenoid, 25
solid state camera, 31
sonar, 7
Spectrum (Timex 2000) computer, 46
speech synthesizer, 10
speed of, 8, 24, 38
spherical robots, 22
Spine robot, 23
steering, 6
switches, 28, 30, 39, 40

teach pendant, 20, 21
"thinking", 34
time of flight recordings, 32
Timex (Sinclair) computers, 46
tinning, 41, 43
tools, 26, 27
tracks, 6, 7, 11
transformer, 8, 16, 39, 43, 44
transistors, 39, 43, 46
Turtle, 16, 30, 38
TV cameras, 3, 11, 13, 14, 28, 29, 35
two-armed robot, 37

ultrasonic sensor, 32
underwater robots, 13

Veroboard, 39, 40, 42
VIC 20, 46
Viking 1, 15
vision, 31, 35
voice synthesizer, 13
Voyager 1, 15

waist, 8
walking robots, 11, 37
welding, 3, 18
"wetting", 41
wheels, 6, 7, 11, 38, 39
wiring diagram, 46
working envelope, 22, 23
wrist, 8, 9, 14, 26, 27, 32, 37

XYZ robots, 22

yaw, 9

Zeaker, 21
ZX81 (Timex 1000) computer, 46

The publishers would like to thank the following for their help:
Artur Fischer (UK) Ltd.
Andrew Lennard, Colne Robotics Co. Ltd.
John Jessop, Jessop Microelectronics Ltd.
Peter Mathews, Upperdata Ltd.

Dr George Russell, Heriot-Watt University.
Milton Bradley Europe.
Economatics Ltd.

The name Usborne and the device are Trade Marks of Usborne Publishing Ltd.

PRINTED IN BELGIUM BY
proost
INTERNATIONAL BOOK PRODUCTION